UNITED STATES
DEPARTMENT OF THE INTERIOR
HAROLD L. ICKES, Secretary
GEOLOGICAL SURVEY
W. C. MENDENHALL, Director

---0---

CIRCULAR 7

---0---

GOLD - QUARTZ VEINS SOUTH OF LIBBY, MONTANA

BY

RUSSELL GIBSON

WASHINGTON

1933

Contents

Page

Abstract ---------- 1
Introduction---------- 2
General geography and geology---------- 2
    Sedimentary rocks---------- 5
    Intrusive rocks---------- 7
    Glacial deposits---------- 8
    Structure---------- 8
Ore deposits---------- 9
    History and production---------- 9
    Occurrence and character---------- 9
    Mineralogy---------- 10
    Tenor---------- 12
    Origin and geologic distribution---------- 13
Mines and prospects---------- 15
    Betty Mae---------- 15
    Blue Bird and Maybe---------- 16
    Diamond John---------- 16
    Fisher Creek---------- 16
    Golden West Mining Company---------- 19
        Little Annie---------- 19
        New---------- 20
    Libby---------- 20
    "Dike" on Libby Creek---------- 21
    Midas---------- 21
    Montezuma---------- 22
    Mustang---------- 23
    Olsen & Switzer---------- 23
    Tip Top---------- 24
    Williams---------- 25

Illustrations

Figure 1 Index map showing location of Libby quadrangle, Montana---------- 3
Figure 2 Map of Southeastern part of Libby quadrangle---------- 4
Figure 3 Plan of Fisher Creek mine---------- 18

Gold quartz veins south of Libby, Montana

By Russell Gibson

Abstract

The area described in this paper is one of rugged mountains, composed chiefly of folded and faulted sedimentary rocks of the Belt series. In the nearby region the beds are intruded by diorite sills and granitic stocks. Glacial drift, including silt deposited in a lake, covers much of the lower slopes of the mountains and partly fills the valleys.

Incomplete records show a production of $200,000 in gold, mostly from the oxidized free-milling parts of quartz veins. Except a few in the Wallace formation the veins are found in sandstone and shale of the Prichard formation. They partly replace the country rock along fractures that commonly follow bedding planes but also cut across them at moderate angles. The veins average between 1 and 2 feet in thickness and reach a maximum of 6 feet. They are cut by faults of small displacement, with which the ore bodies are commonly associated.

Quartz forms 95 percent or more of the veins. Of the remainder, one or more of the minerals pyrite, galena, sphalerite, and pyrrhotite are relatively abundant, and chalcopyrite, arsenopyrite, tetrahedrite, and scheelite are present sparingly. A little silver is present, and in some veins it exceeds the gold considerably.

In many veins an original quartz filling has been slightly brecciated or sheared by movement parallel to the walls. Sulphides and gold have been introduced along the shear planes, partly replacing the quartz and giving the vein a "ribbon" structure. The gold is closely associated with the sphalerite, and silver is more plentiful in the veins that carry galena.

As a rule oxidation has not extended far below the surface, and its chief product is limonite. Cerusite and anglesite are present in some veins, and a little malachite, azurite, pyromorphite, or manganese oxide occurs in a few. Native gold is present in both the oxidized and unoxidized ores.

Although mining of these veins in the past may have been unprofitable, that result was due partly, at least, to difficulties of transportation and milling. With improvement in roads and methods, not to mention appreciation in the value of gold, the deposits are believed to have a promising future.

Introduction

In the course of the areal geologic mapping of the Libby quadrangle, in northwestern Montana, begun in 1929 by the United States Geological Survey, the gold quartz veins in the southeastern part of the quadrangle were examined by the writer during parts of the seasons of 1931 and 1932. Recent mining activity and greatly increased interest in gold deposits generally make it seem desirable to publish a description of the deposits in advance of the more comprehensive report on the quadrangle as a whole.

After his reconnaissance of the region in 1905 MacDonald[1] briefly described some of the gold veins, and a few are mentioned in a press bulletin issued by the Survey in 1931.

Libby (fig.1) is on the Kootenai River in the Cabinet Mountains 24 miles east of the Idaho line. Libby and Troy, the largest towns of the area, are on the main line of the Great Northern Railway. The deposits to be described are between 20 and 25 miles in a straight line south of Libby, in an area called by MacDonald the Cabinet district.

Although the survey of the Libby quadrangle, of which the area described herein is a part, is being made by the United States Geological Survey, Lincoln County, Mont., has cooperated in this survey, and Harvard University has contributed generously from research and scholarship funds.

Prospectors, officers of mining companies, and others residing in the region were unfailing in their cooperation. The writer was assisted in the field at various times by Ian Campbell, A.W. Waldo, R.H. Svendsen, and J.H.Moses.

General geography and geology

The area of gold quartz veins (fig.2) is in the southeastern part of the Cabinet Mountains, northeast of the Clark Fork of the Columbia River. Libby, the largest town of the district, is about 35 miles by road north of the area. Altitudes range from 3,200 to 7,925 feet. The ridges are rugged and their flanks are gashed by deep, narrow gorges. Many of the streams have their sources in small lakes that occupy bowl-shaped basins or cirques hollowed out by glaciers. Enterprising miners have utilized these mountain streams to run mining and milling machinery. The region is thickly forested, and timber suitable for use in mining is abundant.

1/ Calkins, F. C.,and MacDonald, D. F., A geological reconnaissance in northern Idaho and northwestern Montana: U. S. Geol. Survey Bull.384, p. 105, 1909.

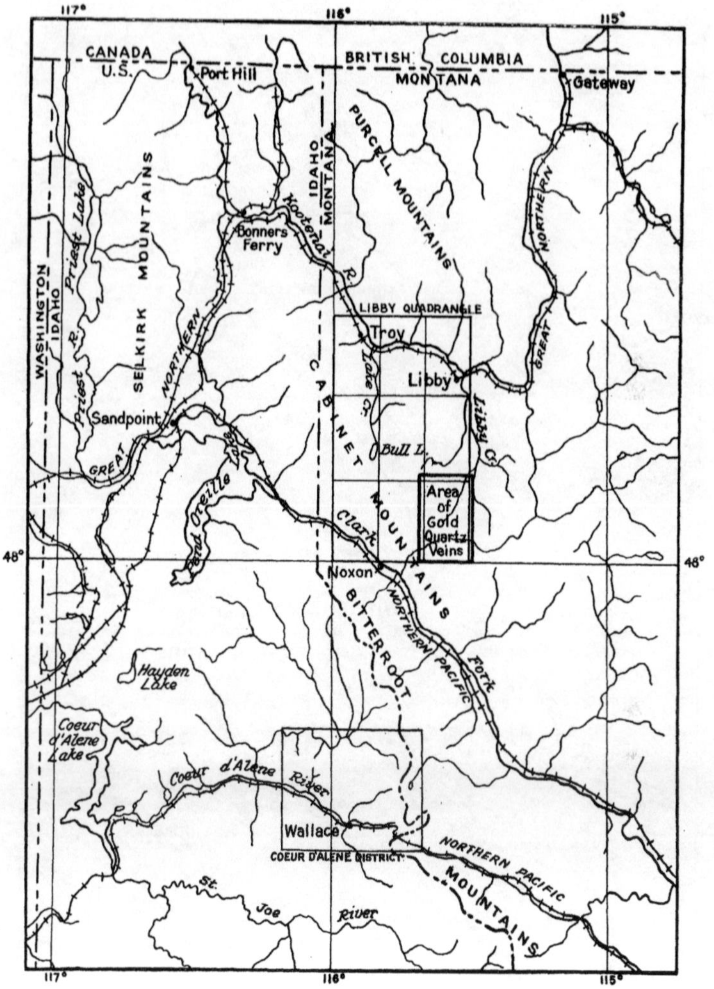

INDEX MAP
SHOWING LOCATION OF LIBBY QUADRANGLE, MONT.
AND AREA OF GOLD-QUARTZ VEINS

Scale
10 0 10 20 30 Miles

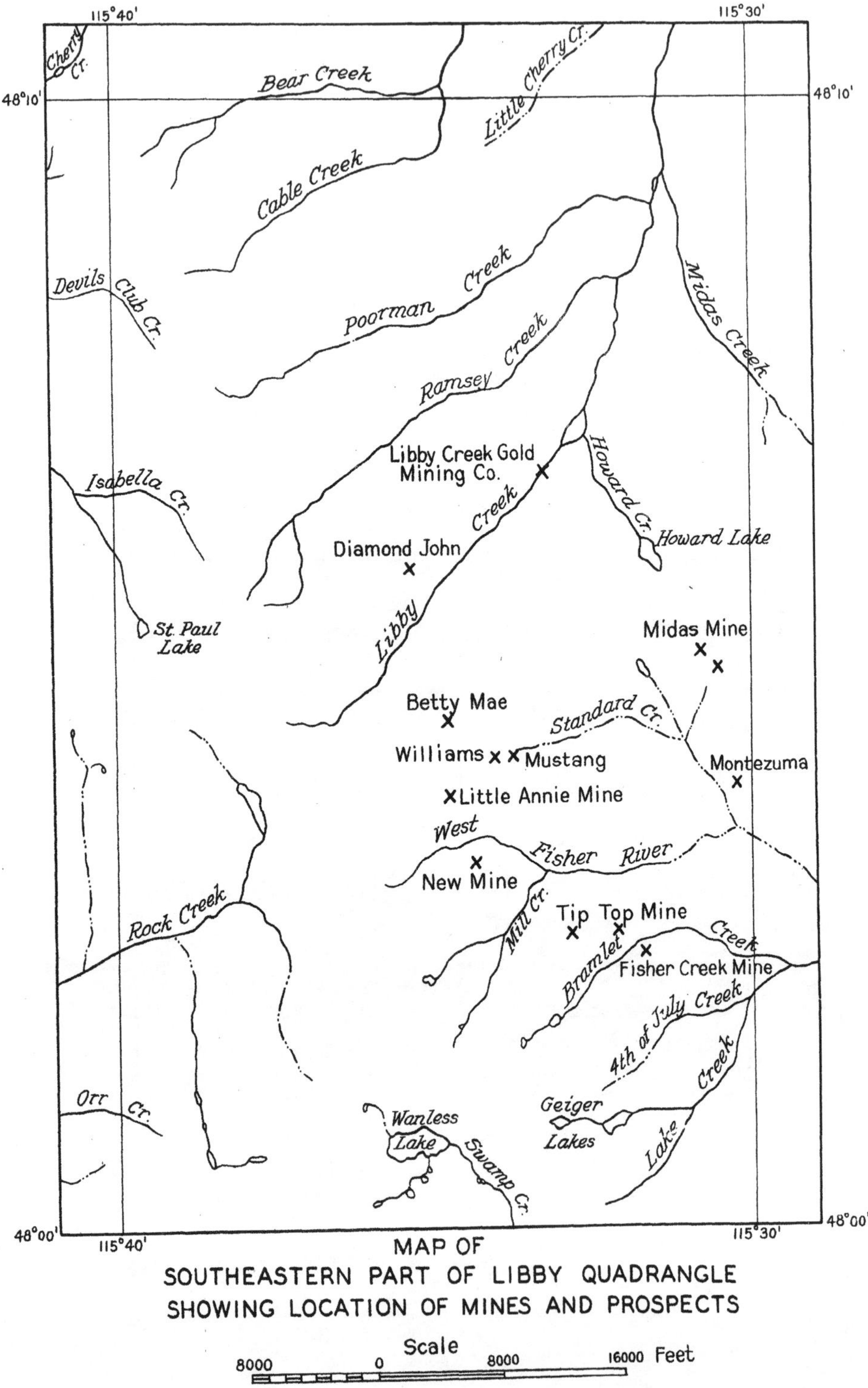

MAP OF
SOUTHEASTERN PART OF LIBBY QUADRANGLE
SHOWING LOCATION OF MINES AND PROSPECTS

The rocks of the Libby quadrangle include Algonkian sediments, which contain the gold-quartz veins, and intrusives of two different ages. The intrusives are of very small areal extent and crop out chiefly in the northern and western parts of the quadrangle. Pleistocene gravel and silt fill some of the valleys and cover the lower slopes of the mountains. The gravel has been worked by placer mining along Libby, Howard, Cherry, and other creeks.

Sedimentary rocks

The dominant rocks of the quadrangle are shales, argillites, sandstones, and quartzites of the Algonkian Belt series. Some of the shales and sandstones contain considerable lime and magnesium carbonate and grade into impure limestones. As the sediments of the Coeur d'Alene mining district, 24 miles south of the south boundary of the Libby quadrangle, belong to this same series, Calkins 2/ used practically the same formation names in his reconnaissance of northwestern Montana and northeastern Idaho in 1905 as he used in the Coeur d'Alene. From oldest to youngest the Belt formations exposed in the area are Prichard, Ravalli (including equivalents of Burke, Revett, and St. Regis), Wallace, and Striped Peak. Above the Striped Peak are several thousand feet of shales and sandstones which Calkins tentatively correlated with the Camp Creek group. In western Montana Calkins used the term Newland instead of Wallace for strata younger than the St. Regis and older than the Striped Peak, but Clapp and Deiss 3/ in 1931 concluded that only the lower part of the Wallace is to be correlated with the Newland of the Belt Mountains, and Calkins now doubts that any part of the Wallace is equivalent to the Newland. Hence the name Wallace is used in this paper for strata that appear to be the equivalent of the Wallace of the Coeur d'Alene section.

The gold quartz veins are in the extreme southeastern part of the quadrangle, where the areal mapping has not been completed, but from the present examination it appears that most of them are in the Prichard formation, though they are by no means confined to that rock. Ransome 4/

2/ Calkins, F. C., and MacDonald, D. F., op. cit., p. 35

3/ Clapp, C. H., and Deiss, C. F., Correlations of Montana Algonkian formations: Geol. Soc. America Bull., vol. 42, p.693, 1931.

4/ Ransome, F. L., and Calkins, F. C., Geology and ore deposits of the Coeur d'Alene district, Idaho: U. S. Geol. Survey Prof. Paper 62, p.141, 1908.

found that all the gold quartz veins exploited in the Coeur d'Alene region up to the time of his examination were in the Prichard formation. Umpleby and Jones 5/ mention one mine, the Alhambra, in the Wallace. In the Clark Fork district of Idaho the gold deposits described by Anderson 6/ are either in quartz diorite sills, lamprophyre stocks, or the Prichard formation.

The Prichard formation consists of dark-colored shale, light-colored fine-grained sandstone, and rocks intermediate between the two. The shale is prevailingly bluish gray but weathers to a rust color. The sandstone is lighter in color, and some beds are nearly pure white. In places the shale is exceedingly thin-bedded and shows alternate dark and light laminae. The base of the formation has not been seen.

The Ravalli, which in the Coeur d'Alene district is a group, divisible into the Burke, Revett, and St. Regis formations, overlies the Prichard and consists of shale, sandstone, and quartzite, ranging in color from white to tones of gray, green, and to a less extent red. The shale is thin-bedded and siliceous and grades into argillaceous and sericitic sandstone that commonly occurs in thicker beds than the shale. Interbedded with the shales are sandstones and quartzites, some of which are slightly calcareous. The middle portion of the Ravalli is dominantly hard white quartzite in beds 1 to 3 feet or more in thickness. The estimated thickness of the formation is 10,000 feet.

The Wallace formation, which overlies the Ravalli, is more calcareous or dolomitic than other formations of the Belt series. It consists of shale and limestone and to a less extent calcareous sandstone and quartzite. The shale is thin-bedded, siliceous, calcareous, and in places ferruginous. A few beds are sericitic. Tones of green and gray predominate, but some beds are nearly white and some are deep red or brownish. Some of the greenish-gray beds weather buff.

5/ Umpleby, J. B., and Jones, E. L., Jr., Geology and ore deposits of Shoshone County, Idaho: U. S. Geol. Survey Bull. 732, p. 125, 1923.

6/ Anderson, A. L., Geology and ore deposits of the Clark Fork district, Idaho: Idaho Bur. Mines and Geology Bull. 12, pp. 123-128, 1930.

The limestone, which is of many degrees of purity, exhibits the same variety of colors as the shale and a great variety of shades of these colors. A noteworthy feature of the Wallace and one that is useful in the identification of that formation is the contorted appearance of some of the limestones. Wavy or contorted patches weather in a cellular fashion unlike any of the dominantly siliceous or argillaceous beds. The Wallace formation is at least 9,000 feet thick.

The beds immediately above the Wallace are provisionally regarded as Striped Peak. They are composed chiefly of hard, even-bedded deep-red and purplish quartzite and sandstone, many beds of which are sericitic. Less abundant are greenish sandstone and quartzite and grayish-green and red impure shale. For about 2,000 feet above the base the beds are prevailingly red, and the top of the Striped Peak is here considered to be the top of the red beds.

Belt sediments younger than the Striped Peak comprise gray, greenish-gray, deep-purple, and bluish impure shales, some of which show alternate dark and light thin laminae; gray, greenish, brown, and purple hard quartzitic sandstones; and a little white limestone. These beds, together with those called Striped Peak, were tentatively referred by Calkins 7/ to the Camp Creek group. Thus far, no exposures of beds above the Wallace formation have been found in the area of gold quartz veins in the southeastern part of the quadrangle.

Intrusive rocks

Intruded at a few horizons conformably with the bedding are sills of dark-colored fine-grained to medium-grained igneous rocks which have been much altered but were originally similar in composition to diorite or gabbro. They range from 50 to 1,000 feet in thickness. These sills have been folded and faulted with the sedimentary rocks. They are more abundant in the northern part of the quadrangle, where they have been prospected to some extent for lead-silver ores, but no gold quartz veins have been reported in them.

In the northern and western parts of the Libby quadrangle, 15 to 20 miles from the area of gold quartz veins, stocks of quartz monzonite, granodiorite, syenite, and similar rocks and dark-colored, commonly greenish granular dikes have been intruded into the Belt series. The dikes are thoroughly altered and consist chiefly of chlorite, biotite, sericite, carbonate, and quartz. These stocks and dikes are presumably to be correlated with intrusives of the Laramide revolution (Cretaceous and early Tertiary), although there is no local means of determining their age. Ore deposits in many districts throughout the West are believed to be related to similar stocks, but no stocks or dikes of these rocks have been found in the area of the gold quartz veins here described.

7/ Calkins, F. C., and MacDonald, D. F., A geological reconnaissance in northern Idaho and northwestern Montana: U. S. Geol. Survey Bull. 384, p. 39, 1909.

Glacial deposits

During the Pleistocene epoch much of the quadrangle was covered by glaciers, which left deposits of ground moraine in all the wide valleys. Remnants of light-colored silt, in places as much as 300 feet thick, form terraces at an altitude of about 2,500 feet along the Kootenai River, Libby Creek, and Lake Creek. This material was deposited in a glacial lake. Some of the gold placer deposits along the present streams appear to have resulted from the reworking of the ground moraine.

Structure

The sedimentary rocks and their included sills have been folded into large open anticlines and synclines that trend north-northwest and plunge toward the northwest. The axial planes of the folds are commonly inclined toward the east. In addition the beds have been sheared and faulted. In many places these movements have followed the bedding; elsewhere they have crossed it. Some of the crosscutting faults are persistent, dip steeply, and show great vertical displacement. It is in or near these faults that some of the largest lead-silver or lead-zinc-silver deposits in the northern part of the quadrangle have been found.

Both bedding faults and crosscutting faults are present in the area of gold quartz veins, but the crosscutting faults are not so conspicuous as in the northern part of the quadrangle. They commonly strike northwest and dip steeply southwest or are vertical. In some places they strike with the beds but dip at a steeper angle. In a few places where these faults are well shown in mine workings they contain as much as 2 feet of gouge and breccia, and the apparent vertical displacement is only a foot or two. The horizontal displacement, however, may be much greater.

In short, the deformation during and after folding involved the breaking of the rocks in the area of gold quartz veins into blocks bounded by bedding and crosscutting fractures, and the squeezing, sliding, and twisting of these blocks. Very few single fractures extend far in any direction, but the two sets of fractures form a system that is continuous over a great area and to a great depth, thus furnishing channels for siliceous and metalliferous solutions from some deep and perhaps remote source.

Ore Deposits

History and Production

Placer gold is reported to have been discovered in this region as early as 1867, and lode mines have been worked for about 50 years. According to the best statistics available, in "Mineral Resources of the United States", a total of $95,000 in placer gold has been reported from the Libby quadrangle since 1901, but it is reasonably certain that this does not include all the placer gold which the district has produced since that date, and no figures are available for earlier production.

According to the records in the volumes of "Mineral Resources" and statistics furnished by mining companies, the value of the total production of all metals from the Libby district since 1901 is more than $6,000,000, but this includes no records of production from lode mines before 1907. The greatest period of activity in the gold-quartz mines southeast of Libby was before 1907; hence only very rough estimates of the gold produced by these properties are available. Estimates of the production from individual mines where it could be ascertained amount in round figures to $200,000. MacDonald reports that before his visit the veins had been sufficiently productive to lead to the construction of three stamp mills. As most of the properties were abandoned in 1905, MacDonald inferred that the ores failed to yield profits below the zone of oxidation. Most of the ore seen in mines being worked during the summers of 1931 and 1932 and remnants of ore in abandoned prospects showed some effects of oxidation. In some veins which, according to assay, contain both gold and silver no trace of sulphides remained. In the Little Annie, on the other hand, rich ore containing visible gold was only slightly oxidized, and the gold was associated with sphalerite and galena.

Occurrence and character

The gold quartz veins of the Libby district occur, for the most part in sandstones and shales of the Prichard formation. So far as known to the writer the Midas is the only mine and the Montezuma the only large prospect whose ores are in the Wallace formation. No veins in the sills or dikes have been sufficiently rich in gold to be worked chiefly for that metal.

The veins are commonly parallel to the bedding and have therefore been called bed veins. Some of these veins, however, cut across the beds at a low angle; and some distinctly crosscutting veins change their course and become bed veins. The veins pinch and swell within short distances, or one vein may split into two or more branches that fray out into the country rock. Other veins come in above or below a pinched vein, or, if one vein splits into several, the intervening country rock may be mineralized. Where the veins are not parallel to the bedding, they commonly cut across at low angles and invariably have a steeper dip than the beds, whether the beds are comparatively horizontal ("flat") or not. It is said that in the vicinity of the Little Annie mine, where there are "flat" bed veins and steeply dipping crosscutting veins, the bed veins are much richer in gold than the crosscutting veins.

The veins range in thickness from a few inches to 6 feet and average between 1 and 2 feet. Inclusions of country rock in all stages of replacement by quartz are common, and where a vein is several feet thick such inclusions make up a considerable portion of it. These inclusions have had no apparent effect on the introduction of metals.

It is common to find evidence of mineralization along bedding planes related to faults and shear zones of comparatively steep dip. Good ore has been mined from bed veins in the Little Annie, the Tip Top, and the Fisher Creek mines near steeply dipping faults. In the Little Annie a little rich ore was mined from a vein in a vertical fault. At the Midas the relations are not clear, but there is some evidence to indicate that the vein occupies a shear zone. At the Blue Bird prospect a crosscutting shear zone containing in places as much as 2 feet of quartz has been traced on the surface for 1,500 feet.

Mineralogy

From 95 to 98 percent of most of the veins is quartz. It is milky white, where it is not stained as the result of oxidation of sulphides, and rarely shows vugs. Under the microscope its grains are seen to range from 0,005 to 17.00 millimeters in greatest dimension, though the average is only about 1.0 millimeter. The grains very commonly interlock somewhat like the parts of a jigsaw puzzle and rarely show crystal faces. As a result of shearing, the quartz in many places shows ribbon or platy texture parallel with the bedding, and the beds adjacent to the veins are sheeted in the same direction. This is commonly developed in good ore but may be present also in poor ore. Under the microscope the ribbon quartz shows strain shadows, brecciation, and faulted portions of small veins. Sericite, sulphides, and gold are related to these brecciated areas--sericite less strikingly so than the metallic minerals. Very commonly both gold and sulphides are seen under the microscope to have replaced the small brecciated grains of quartz, but both gold and sulphides have also penetrated and replaced large solid grains of quartz.

Sericite, the only other conspicuous gangue mineral, is present in all the veins. Apparently it is residual from incompletely replaced country rock and therefore is not, strictly, one of the vein minerals. Small amounts of siderite, dolomite, and calcite are present, as well as minute quantities of chlorite and epidote, which have probably been derived from the country rock.

Pyrite, galena, sphalerite, and pyrrhotite are the most abundant of the metallic minerals, though commonly not more than one of these is abundant in any one vein. Chalcopyrite is present in about two thirds of the veins seen but is everywhere scarce. Arsenopyrite, tetrahedrite, and scheelite likewise are scarce.

Pyrrhotite, pyrite, and sphalerite are early minerals, deposited in the order named. In the veins they are commonly coarse-grained, but where phyrrhotite and pyrite are disseminated in the wall rock near the vein they are fine-grained. Sphalerite is commonly confined to the veins. Pyrite is the most widespread of the three. In a few places these sulphides had been shattered before they were replaced by others, but the shattering of the sulphides is much less common than the brecciation of the quartz.

Galena and chalcopyrite follow sphalerite in sequence. Galena is as widespread as pyrite and a little more abundant. Specimens high in galena have yielded the best assays for silver.

Native gold is coarse enough to be seen easily in ores from the four largest producers described in this paper, and the following description of its relations to other minerals is based on examination of specimens from those four properties---the Fisher Creek, Golden West, Midas, and Tip Top.

Gold appears to be concentrated in those veins or parts of veins that contain or have contained sulphides. Thus far, the richest ore has come from quartz veins at least 1 foot thick. Where gold is reported to be disseminated through several feet of sedimentary rock, it is found on close examination that the rock is partly replaced by tiny quartz veins which contain sulphides and thus transform the rock into a low-grade lode. Gold is also present alone in quartz, most commonly in those portions of the vein that have been sheared and brecciated. Next to quartz, the commonest associate of gold is sphalerite. Under the microscope gold is seen in many specimens to have replaced sphalerite or to occur as a replacement mineral along boundaries between sphalerite and some other mineral, commonly quartz, galena, pyrite, or pyrrhotite. Gold was found less commonly at grain boundaries of galena, pyrite, or pyrrhotite and quartz. A little sericite, which may be residual, is seen in the brecciated quartz areas with the gold and sulphides. Small amounts of tetrahedrite and scheelite are present in the Midas ore, but neither mineral was seen in close association with gold.

Where gold replaces sulphides, it is smooth and dark yellow. The grains are shapeless and irregular or long and wirelike or lenticular. Where gold has replaced sphalerite, it may exhibit straight edges controlled by the sphalerite cleavage. . In oxidized ore the gold is gnarled, rough, or spongelike.

During or after the deposition of gold a small amount of fine-grained pyrite and more quartz were added. In vuggy ore pyrite of this stage can be seen in very small crystals perched on quartz; in massive ore it appears as strings of small cubes replacing early minerals.

Recent glaciation has removed the upper parts of the veins, and erosion since glaciation has been rapid. Oxidation, therefore, is not conspicuous, but its extent varies greatly from place to place--for example, very little sulphide is left unoxidized in ore mined at the Tip Top, whereas ore at the Little Annie, which is not much deeper than the Tip Top, shows scant oxidation. Visible native gold is common in both ores. There are all gradations of oxidation between these two. Limonite is the most common oxidation product, cerusite and anglesite are present in some ores, and scant malachite, azurite, pyromorphite, and manganese oxides were seen in a few places. A little native copper was seen at the Betty Mae.

Tenor

Mill recoveries reported by the operators for some of the mines range from $6 to $50 a ton. 8/ A shipment of 39 tons of high-grade ore reported from one mine yielded nearly 4 ounces of gold and about 1 ounce of silver to the ton. Assays of selected samples of ore from still other properties yielded from a trace to 24.86 ounces of gold to the ton. No other metal is sufficiently abundant in the gold quartz veins to pay for mining it alone, but silver, lead, copper, and tungsten make up part of the total production.

Early records are not complete enough and well-authenticated assays are too few to establish any ratio between gold and silver that would be significant. According to smelter returns and assays of samples selected by the writer, silver is by weight more abundant than gold except in the very richest ores. Three selected samples of the richest ores, which yielded on assay 1.66, 4.91, and 24.86 ounces of gold to the ton, contained more gold than silver.

8/ All values of gold given in this paper have been computed on the basis of $20.67 a fine ounce.

Origin and geologic distribution

Gold quartz veins similar to those of the Libby quadrangle have been found in other mining districts where they are clearly related to intrusive granitic rocks, but the nearest known outcrop of a deep-seated intrusive to which the ores of the Libby quadrangle may be related is 13 miles to the northwest. It is a granodiorite stock about 20 square miles in surface area. There are no gold quartz veins in it, but several lead-silver prospects and at least two mines, the Snowshoe and the Glacier, are within a radius of 5 miles of the stock. It is not an unwarranted assumption that an offshoot from this stock or some other related intrusive may lie a short distance below the surface in the area of the gold quartz veins, but no contact metamorphism or zoning within the area has been observed thus far to bear out this suggestion.

There are more gold quartz mines and prospects in the Prichard than in any other formation in the Libby quadrangle, but no special beds of the Prichard are more favorable than any other. Fairly pure gray shale which weathers to a rust color, finely laminated siliceous shale, argillaceous sandstone, and white sericitic sandstone are the wall rocks seen in different properties. At the Midas, the largest active mine in the summer of 1933, a vein was being worked in impure sandstone, calcareous and sandy shales, and impure limestone of the Wallace formation. Evidently the kind of wall rock did not influence the deposition of the quartz veins. Referring to gold quartz veins of the Coeur d'Alene district, most of which are in the Prichard formation, Ransome 9/ says that "it would be difficult to disprove the suggestion that the occurrence of the gold ores in the Prichard is merely a coincidence, the slates happening to be exposed by erosion over areas where the deposition of gold had been determined by deep-seated causes entirely independent of the rocks now visible at the surface."

It is the opinion of some of the prospectors that only blanket veins--that is, essentially horizontal bed veins--are likely to be profitable gold properties, but so far as the writer has observed the attitude of the country rock had little to do with the deposition of the ore. The productive bed veins in the Little Annie mine are nearly horizontal. The Midas vein has an average dip of 53°, and its country rock has the same or nearly the same attitude. The Fisher Creek ("Branagan") bed veins have dips between these extremes. The Little Annie mine is on the crest of an anticline, the Tip Top mine is not far east of the crest, and the Fisher Creek, Midas, and Montezuma mines are on the northeast limb of another anticline.

9/ Ransome, F. L., and Calkins, F. C., Geology and ore deposits of the Coeur d'Alene district, Idaho: U. S. Geol. Survey Prof. Paper 62, p. 143, 1908.

Although cross faults with steep dip are not so conspicuous in this area as in the northern part of the quadrangle, and it is commonly reported that veins in them are not so rich as the bed veins, some of the good shoots in the bed veins are closely associated with cross faults. In the Little Annie and Tip Top mines there are vertical faults of small net displacement near which good ore has been mined; in the Fisher Creek mine steeply dipping or vertical faults are present in most of the old stopes, and rich assays were obtained from remnants of ores in these faults; and in the Midas mine the entire Midas vein evidently occupies a fault or shear zone. Along some of the cross faults bed veins have been clearly displaced, but it is not everywhere evident whether the displacement occurred before or after the deposition of the sulphides and gold. In some places, as at the Fisher Creek mine, the shattering of sulphides indicates renewed movement along faults some of which probably antedate any vein deposition, whereas movement along others may have occurred only after the deposition of quartz. Good ore is locally associated with faults that may have originated since the deposition of sulphides and gold, but this association is evidently accidental.

Movement, therefore, has taken place along these cross faults during at least three different stages--prior to or during the introduction of quartz, subsequent to the introduction of quartz, and subsequent to the introduction of sulphides.

Movement has taken place also within the bed veins parallel with the walls and has developed sheeting or ribbon texture in the vein. Under the microscope, thin and polished sections of the ores show that solutions found brecciated areas in these sheeted zones favorable places for the deposition of sulphides and gold.

The features above set forth are similar to those in gold quartz veins of certain other districts. In the Alleghany district, California, according to Ferguson and Gannett 10/ certain high-grade shoots in gold quartz veins are associated with minor transverse faults that were formed after the deposition of the quartz, and other shoots are localized where the shattering of quartz by shearing along and in the vein permitted ingress of the gold-bearing solutions.

In comparing the gold lodes of the Sumpter quadrangle with those in similar districts, Hewett 11/ has called attention to the trend of interpretation of the relations of quartz, sulphides, and gold in veins in many

---

10/ Ferguson, H. G., and Gannett, R. W., Gold quartz veins of the Alleghany district, Calif.: Am. Inst. Min. Met. Eng. Tech. Pub. 211, pp. 35-36, 1929; U. S. Geol. Survey Prof. Paper 172, pp. 56-58, 1932.

11/ Hewett, D. F., Zonal relations of the lodes of the Sumpter quadrangle [Oregon]: Am. Inst. Min. Met. Eng. Trans., 1931, pp. 305-346.

districts. Evidence shows "that the veins were first filled with quartz that was barren of gold and sulphides and that these minerals were deposited late in the history of the vein, with or without associated quartz, after one or more intervening epochs of crushing."

Mining conditions

For about 30 years mining operations in this area have been rather spasmodic and without much profit. Among the reasons assigned are lack of capital, remoteness of location, and certain inherent features of the deposits, such as the thinness of the veins, their lenticular character, and their attitude. Libby, the nearest railway point from which supplies must be brought, is 35 miles to the north and is connected over part of the distance by poor roads. The former manager of the Fisher Creek mine estimates that his treatment losses were exceedingly high, perhaps as much as 40 percent. Other handicaps that have meant losses in the past are lack of adequate machinery such as power drills, shortage of water for power, and, for some of the mines, the expense of pack-horse transportation. In 1932 the Midas was the only property in the district that had electric power and modern milling equipment.

As many of the veins are thin and not continuous, much waste rock must be mined, and no large bodies have thus far been developed that would justify an effort at extensive low-cost operation; on the other hand, with improvement in transportation and appreciation in the value of gold many of the veins promise a profit if operated on a sufficiently modest scale. Records of assays quoted by owners of the Tip Top mine suggest that there may be in this area bodies of low-grade gold ore that consist of several quartz veins and intervening country rock impregnated by sulphides, but thorough sampling would be necessary to determine whether any such bodies are sufficiently extensive and continuous to justify large-scale, low-cost mining.

Mines and prospects

Betty Mae

The Betty Mae group comprises six unpatented claims near the head of Goat Creek, a tributary of Libby Creek near its source. An excellent cabin has been erected on the property. Six shallow openings have been made on relatively horizontal quartz veins in northwestward-dipping Prichard sandstones and sandy shales. The chief primary metallic minerals are pyrrhotite, sphalerite, chalcopyrite, and galena. As a result of oxidation, small amounts of limonite, native copper, melanterite, anglesite, and cerusite have been developed. The vein quartz was sheared before sulphide deposition, and movement in the same veins afterward has brecciated the sulphides to a slight extent. According to Frank Warington, one of the owners of the property, a selected sample yielded 10 percent of copper and 20 ounces of silver and half an ounce of gold to the ton.

Blue Bird and Maybe

The Blue Bird and Maybe claims are on the west side of Hoodoo Creek east of the eastern border of the quadrangle. Several shallow open cuts have been made on a shear zone 2 feet wide in greenish and gray shales that strike N. 8° E. and dip 34° SE. The shear zone, which strikes N. 50° W. and dips 78° SW., has been traced for about 1,500 feet on the surface. In places it shows as much as 2 feet of quartz, which contains patches of galena and chalcopyrite. Oxidation of this part of the vein has produced limonite, malachite, azurite, anglesite, and pyromorphite. According to Herman Bockman, who owns the claims, assays of a picked sample that contained a very high percentage of copper yielded $2 in gold to the ton.

Diamond John

The Diamond John prospect is on the north side of Libby Creek at an altitude of 4,350 feet, about 3 miles by secondary road and trail above the main road to Howard Lake. A 60-foot adit has been driven N. 25° W. along two westward-dipping bed veins 3 feet apart, each of which shows a maximum of 16 inches of quartz. Pyrrhotite, the dominant sulphide, and small amounts of galena, sphalerite, pyrite, and chalcopyrite are related to sheeted and shattered areas in the quartz and are especially abundant where the sheeting is parallel with the bedding. The Prichard wall rock is light brownish-gray sandstone and bluish-gray siliceous shale. The oxidation of disseminated sulphides in these rocks stains the weathered surface brown.

According to Isak Ludvikson, one of the owners, assays of selected samples show small amounts of both gold and silver.

Fisher Creek

The Fisher Creek Mining Co. owns six lode claims and two placer claims, all patented, on the south side of Bramlet Creek a little less than 2 miles above the junction of the West Fisher River and Bramlet Creek roads. The mine, called locally the "Branagan," has not been worked for many years. It is reported that in 1901-3, in a run of 26 months, the Branagan owners "cleaned up" $150,000 and that at that time their milling was only 60 percent efficient.

Quartz bed veins in the eastward-dipping Prichard formation have been opened up at several places along the outcrop, and all the adits are connected underground by the extensive gently dipping stopes. Five of the openings have been numbered on the map (figure 3) for the sake of convenience in referring to them. As the map shows, the stopes are exceedingly irregular in shape and range in height from 2 or 3 feet to 10 feet. The thickest quartz vein seen is 4 feet thick, but wall rock near the larger veins contains small veins or is impregnated with quartz. In adit 1 two 1-foot bed veins are separated by as much as 3 feet of rock but in places the quartz breaks across, from one vein to the other, and the entire 5 feet has been mined. In this same adit a 1-foot vein thickens to 2 feet in a distance of 40 feet. Commonly, if one vein pinches out, another is found above or below it along another bedding plane, and the two veins may overlap.

Sulphides are not common but are present in all the quartz veins and are everywhere more or less oxidized. In decreasing order of abundance the primary metallic minerals seen are pyrite, galena, pyrrhotite, chalcopyrite, sphalerite, and native gold. Pyrite is present in all the ore that was examined, and galena in nearly all. The other sulphides are rare. Gold was seen only in specimens from the face in adit 4. The sulphides are irregularly distributed through the vein in disseminated grains and bunches but appear to be concentrated in or close to sheared areas. Not all such areas in the quartz contain sulphides, however, partly because some of the movement that fractured the quartz took place after the sulphides were formed. As a result of partial oxidation the sulphide patches in all parts of the mine are friable and discolored with small quantities of iron oxide and less commonly manganese oxides. So far as could be determined by the remnants of ore left at the faces of the different adits the oxidized ore is not yet exhausted. Sericite, the only other gangue mineral in addition to quartz, is present in greater or lesser amounts in all specimens. As sericite and quartz are the chief constituents of the wall rock it is not everywhere possible to determine whether these minerals were introduced with the veins or constituted part of the original walls.

Steeply dipping or vertical faults, some of which contain from 12 to 24 inches of breccia and gouge, are conspicuous in the backs or in the walls of the stopes and drifts. They strike north or a little east or west of north, and most of those where the complete attitude could be determined dip west. The fault at the face in adit 1 contains a band of breccia 1 foot thick. The fault in adit 4 near the second crosscut to adit 5 contains a 2-foot band. The fault in adit 2 contains over 1 foot of breccia and gouge. Because of the similarity of the beds above and below the veins, measurement of the amount and direction of displacement is difficult. The apparent vertical displacement is only a foot or two, and the downthrown side is consistently to the east. The horizontal displacement may be greater. Mineralization along the faults has been slight or lacking, and they have not been explored to any extent by the miners. The last movement along a few of the faults where it could be determined with certainty was postquartz. It is significant, perhaps, that these faults are present in every stope and that the strongest, most persistent fault is in the largest stope.

PLAN OF FISHER CREEK MINE

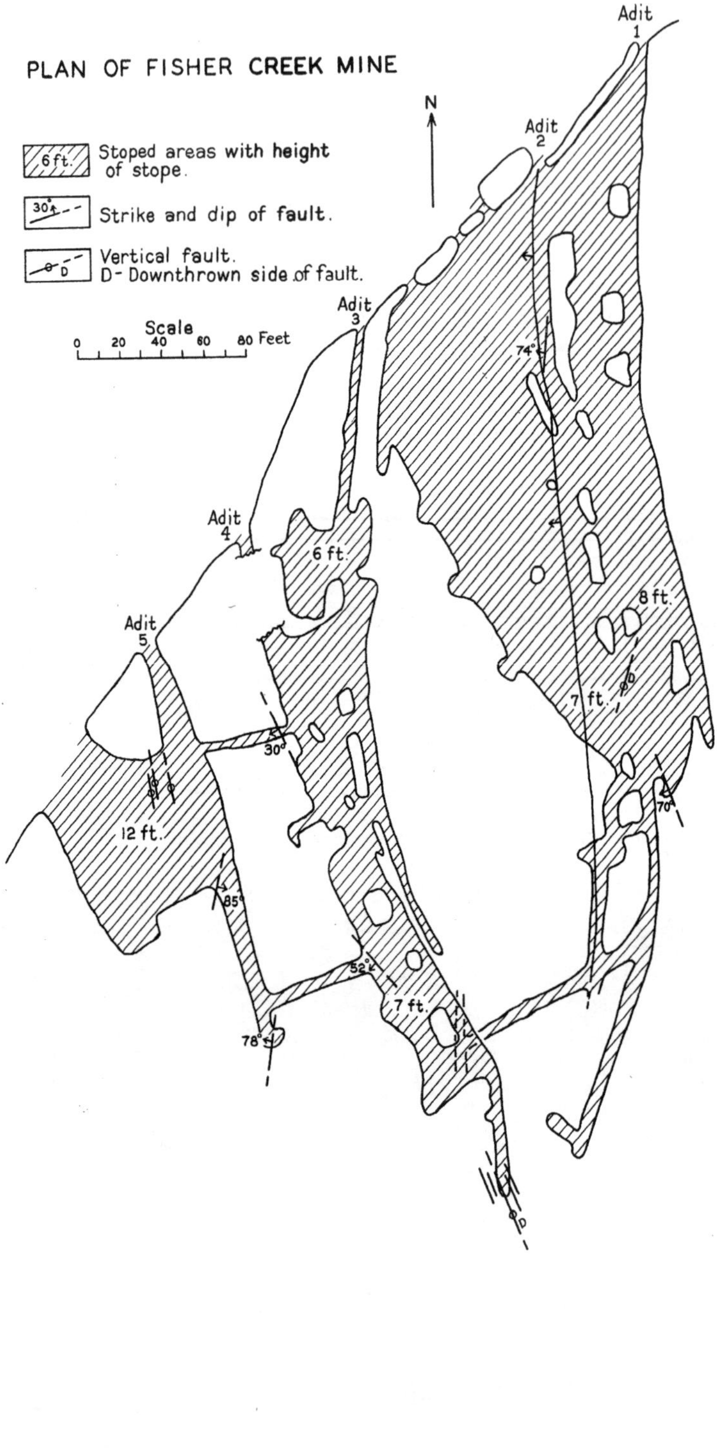

A specimen of incompletely oxidized ore containing very small amounts of sphalerite, galena, and pyrite yielded on assay 1.66 ounces of gold and 1.44 ounces of silver to the ton.

Golden West Mining Co.

The Golden West Mining Co. has two properties on the West Fisher River, the Little Annie and the New mine. The camp may be reached by a secondary road that connects with a good highway to Libby. The equipment includes a Diesel engine and compressor whence air is piped to both mines. The Little Annie is 1,200 feet and the New 700 feet vertically above the camp. Several large cabins have been erected at the central camp in the valley and at the Little Annie mine.

Little Annie.- The Little Annie mine, on the north side of the stream, is worked through two nearly parallel adits which trend N. 10° E. in sandstones and finely laminated sandy shales of the Prichard formation. The shales are bluish gray on fresh fracture but weather to rusty brown. The adits are connected by a crosscut, and at the time of examination a total of 300 feet of workings had been opened.

The ore occurs in quartz veins as much as 2 feet in thickness that are for the most part parallel with the beds, which dip gently to the northwest. The chief primary metallic minerals are sphalerite, pyrite, galena, and native gold, and the gold is sufficiently abundant to be seen easily in many hand specimens. The close association of gold with sulphides in this and other mines and the order of deposition of these minerals indicates that the sulphides influenced the deposition of the gold. A little pyrrhotite and a very little chalcopyrite are present. Small quantities of sericite and siderite are present in the quartz. The wall rock near the vein contains sufficient disseminated pyrrhotite to cause the rock to weather

Faulting of the veins parallel with the beds is shown at the contact by grooved and slickensided quartz surfaces and within the veins by shattered areas in the quartz. Sulphides and gold characteristically follow these shattered areas. This can be seen in the hand specimen but more strikingly under the microscope. Obscure, nearly vertical cross faults were seen which show small net displacement.

The ores being mined at present show scant oxidation. This is of unusual interest in view of the history of several other gold lodes in the Fisher River drainage basin which according to MacDonald 12_/ were not workable below the zone of oxidation.

12_/ MacDonald, D. F., Economic features of northern Idaho and northwestern Montana: U. S. Geol. Survey Bull. 285, p. 50, 1906.

According to officials of the Golden West Mining Co., one shipment of 39 tons of ore from the Little Annie yielded 3.874 ounces of gold and 1.05 ounces of silver to the ton.

On Golden West ground east of the Little Annie there are bed veins as much as 4 feet in thickness with scant sulphides. These veins pinch and swell rapidly. One of the largest of them decreases from 24 to 12 inches within a distance of 8 feet. At this place, however, smaller parallel veins come in above the main vein. The bed veins are crossed by veins half as thick, which themselves seem to turn off along the bedding at no great distance. It is reported that assays show the crosscutting veins to be not nearly so rich as the bed veins.

New.- Very little work has been done on the New mine, on the south side of the West Fisher River. A short adit at an altitude of 4,900 feet has opened up nearly horizontal quartz bed veins similar to those at the Little Annie mine, but they are neither so thick nor so rich in sulphides. Galena, zinc blende, and pyrite are the chief metallic minerals. The ore is partly oxidized, and small quantities of limonite are present. Pannings of this oxidized ore showed native gold. The wall rock is gray sericitic sandstone and gray, finely laminated sericitic siliceous shale. Assays of ore from the New mine are said to average $50 to the ton, chiefly in gold.

Libby

The Libby prospect includes two unpatented claims on the south side of the West Fisher River high on the slope west of the Tip Top group, from which it is reached by trail. Five short adits, the longest 100 feet in length, have been driven in a southerly direction to explore quartz veins. The country rock consists of gray sericitic sandstone and sericitic shale, which probably belong to the Prichard formation. These rocks are commonly stained brown by iron oxide, which is scattered abundantly through some beds in tiny grains. The cleavage of the shale is in many places more prominent than the bedding.

The quartz veins range in thickness from a fraction of an inch to 2 feet or more and are exceedingly irregular--for example, one vein pinches from 12 to 5 inches in a distance of 13 feet, and the 5-inch part then branches into several small veins; but veins are so numerous that another may be picked up above or below. Some veins follow the bedding planes for a short distance and then break across along the rock cleavage to a higher bedding plane. Between two parallel bed veins there are in places numerous links of vein quartz along the rock cleavage. The veins cut across the cleavage also. The irregularity and diminution in size of veins continues down to microscopic dimensions. When studied under the microscope specimens of country rock near the veins are seen to be cut by numerous tiny quartz veins that have gradually replaced sandstone or shale. The only metallic minerals seen in the veins are galena and its oxidation products, cerusite, anglesite, pyromorphite, and a little iron oxide. Former pyrite is completely converted to limonite.

According to John Fredericks, owner of the property, one sample assayed yielded $13 in gold to the ton.

"Dike" on Libby Creek

In the Ravalli formation a group of mineralized beds commonly spoken of as a "dike" crop out on Libby Creek between Libby Creek Falls and the mouth of Howard Creek. There has been some placer mining on Libby Creek below these beds, and they have been regarded as a probable source of the gold. In the summer of 1932 the Libby Creek Gold Mining Co. was placer mining in the vicinity of the "dike" and also indicated its intention of exploring the "dike" itself.

Along the stream bed for a short distance below Libby Creek Falls the Ravalli formation is exposed at several places for about 300 feet. The beds consist of hard arkosic quartzitic sandstone that is stained buff by iron oxide and has an average grain size of 0.20 mm. They dip 44° NE. Along the joints are dendrites of manganese oxide. Under the microscope it is seen that from one tenth to one fourth of some beds is made up of detrital grains of potash feldspar, part of which is altered to sericite. Other fine-grained sericite, formerly clay material, fills interstices between quartz grains. A few well-rounded grains of zircon are present.

The sandstone is cut by small, irregular quartz veins which, before oxidation took place, contained a little sulphide. As in some places the siliceous solutions soaked into the sandstone and replaced other minerals or filled interstices, the quartz veins are not everywhere well defined. Under the microscope tiny veinlets a fraction of a millimeter in width can be discerned. From the distribution of the iron oxide stains it seems clear that most of the sulphide was associated with the introduced quartz.

Some prospectors claim that colors of gold can be panned from some of the sandstone beds, but others disagree. Assays of specimens taken at intervals for a distance of 300 feet across the mineralized sandstone as exposed in the creek bed showed very small amounts of gold and silver.

Midas

The Midas mine, which includes the former Rose Consolidated property, is on the county highway about 1 mile southeast of Howard Lake and 21 miles by air line almost directly south of Libby. It comprises eight unpatented lode claims and one placer claim in the group. According to Mr. Frederic Keffer, engineer for the mine, who gave the writer information about the history, production, and mill equipment, the mine was taken over by the Midas Gold Mining & Milling Co. in 1926 and has since been absorbed by or consolidated with the Spokane-Idaho Copper Co., of Spokane, Wash.

In addition to stopes more than 3,000 feet of drifts, crosscuts, and raises are open and accessible. Diesel engines furnish power for an electric hoist, pump, locomotive, and other machinery. Equipment in the 75-ton mill includes jaw and cone crushers, ball mill, sampler, Dorr classifier, flotation machines, concentrating tables, and cyanide tanks.

The claims were first located in 1905, but there is no record of early production except between 1916 and 1918 when gold bullion and tungsten concentrates amounting to $27,000 were produced. In a test run in 1928 about 1,000 tons of ore, much of it of low grade, netted $6,034. This lot contained 317.14 ounces of gold, 503 ounces of silver, 77 pounds of copper, and 1,029 pounds of lead. In 1929 production was negligible. From August 1, 1932, to July 1, 1933, the ore produced was valued at $20,962 and included 990.79 ounces of gold besides silver, lead, and tungsten. The production for July 1933 is estimated at about $5,000. The minimum total production of the Midas is therefore valued at about $59,000.

The ore occurs in a quartz vein that ranges from a few inches to 6 feet and averages at least 1 foot in thickness. For the most part the vein is parallel with the beds, which here strike N. 20° -30° W. and dip 40°-60° NE. The average dip is about 55°. Locally, the vein may dip more steeply than the bedding and cut across it, but the relations are not perfectly clear. At the Montezuma prospect, 2 miles south of the Midas, similar veins locally dip at a steeper angle than the beds. In places small veins split off and enter the hanging wall, but commonly the vein is clean and well defined. The wall rock is the Wallace formation and consists of calcareous shales, some of which are paper-thin; sandy shales or sandstones, commonly more massive than the shales; and limestones that are banded and in places contorted. Their prevailing color is gray, they weather buff, and the limestones show characteristic cellular forms.

Next to quartz, scheelite and carbonate are the most abundant minerals in the vein. Small amounts of sericite and chlorite are present, and scant galena and tetrahedrite were seen. With the aid of a hand lens, native gold can be seen, associated with iron oxide, malachite, and scheelite. The gold is gnarled and wire-like and deep yellow. Everywhere the ore is partly oxidized, but as sulphides were not abundant the vein quartz and wall rock are not much stained.

Brecciated and silicified hanging-wall rock, in which the matrix of the breccia is vein quartz and the fragments limestone, indicates that faulting took place before or during the introduction of quartz and that the vein may occupy a fault. Faulting along northward or northeastward trending fault planes certainly took place after the formation of the quartz vein. In at least two places these faults offset the vein a little.

Montezuma

The Montezuma group of six claims is about 2 miles south of the Midas mine, on the north side of the West Fisher River just below the mouth of Standard Creek, and is reached by a first-class road from Libby. Quartz veins from a few inches to 20 inches in thickness have been explored by several open cuts and short adits, the longest 385 feet long. The altitude at the portal of this adit is about 3,575 feet. The veins strike northwest, dip 40°-50° NE., and in some places are parallel with the sedimentary wall rocks. Elsewhere the veins dip at a steeper angle than the beds. The veins pinch and swell and have been sheared, commonly parallel with the walls. The wall rocks are light-colored sandstone, buff calcareous shale, and buff-weathering dolomitic limestone of the Wallace formation.

The chief metallic minerals in the veins are tetrahedrite, galena, pyrite, and chalcopyrite. None of these are abundant. As the prospects are shallow, all the ore seen was partly oxidized, and malachite, azurite, and limonite are present. Assays show that partly oxidized ore in which sulphides are scant contains both gold and silver.

Mustang

The Mustang group of four claims is on the south side of Standard Creek about 2 1/2 miles south-west of the Midas mine. The claims are reached by a trail from the road to the Midas. An adit 300 feet long has opened up a quartz vein from 7 to 18 inches thick which in places splits into a lode of similar thickness composed of thin, closely spaced veins. In an open cut on top of the ridge a few hundred feet above the adit is a lode of the same type 30 inches thick which is thought by the owners to be a continuation of the lode in the adit.

The wall rock in the adit comprises impure gray to buff sandstone and shale. Some of the sandstones are calcareous, and all are shaly; the shales contain more or less sand. Tiny disseminated grains of pyrrhotite are so numerous in some of the sandstone beds that a drop of hydrochloric acid placed on a scratched area on a specimen will evolve hydrogen sulphide. One of the sandstone beds is nearly pure white and resembles the poorly cemented sandstone bed on the Tip Top property, on Blacktail Mountain described below, which yields a little gold by panning. The beds strike N. 30° W. and dip 28°-33° NE. The vein strikes with the country rock but dips at a steeper angle, in places as steep as 45°.

As the workings are shallow, all parts of the vein seen are well oxidized. Scant pyrite and galena are present, but the quartz is honeycombed and stained with iron oxide where sulphides have weathered out. A little pyromorphite is present. The vein is sheeted and shows "ribbon quartz" structure. This may be partly the result of replacement of sheared wall rock, as suggested by the fact that the wall rock close to the vein is sheeted parallel to the vein and contains thin veins and podlike masses of quartz. Rock of this type is thoroughly iron-stained and therefore conspicuous. Sulphides and quartz were deposited together in areas made permeable by shearing. After the vein was formed shearing again occurred and brecciated the quartz in places.

A specimen of the sheeted, brecciated, iron-stained vein quartz showing no sulphides yielded on assay 1.49 ounces of gold and 2.11 ounces of silver to the ton.

Olsen & Switzer

Several quartz veins parallel with the bedding have been opened up by short adits and open cuts at altitudes between 4,650 and 5,350 feet near the head of Bramlet Creek, on eight unpatented claims belonging to L. J. Olsen and Elmer Switzer. The country rock is bluish-gray to gray shale and is horizontal or dips very gently to the north, northeast, or southeast. These bed veins range from 6 to 48 inches in thickness and have been faulted commonly parallel with the bedding, as is shown by the gouge and the shearing and grooving of the quartz. The chief sulphide, galena, constitutes about 90 percent of the metallic minerals; small amounts of pyrrhotite, zinc blende, chalcopyrite, and pyrite make up the remainder. The sulphides are related to the sheared and granulated areas in the quartz. The claims are reached by trail from the road to the Tip Top mine. According to Mr. Switzer, assays of samples from one of the veins yielded $5 in gold to the ton.

Tip Top

The Tip Top group, formerly called the Blacktail, includes 2 1/2 patented and 11 unpatented lode claims on the slopes north of Bramlet Creek and south of the West Fisher River, comprising Blacktail Mountain. The property includes a 10-stamp mill on the Bramlet Creek road less than 2 miles above the junction with the West Fisher River road a good highway to Libby. The mill is run by water power from Bramlet Creek and is connected by a small aerial tram with an ore bin higher on the mountain. Ore mined on the north side of the mountain is hauled around to the bin by sled over an old narrow-gage railroad grade. Several cabins have been built near the Bramlet Creek road. The total production of the property is not known, but according to Mr. O. V. Miller, one of the present owners, about $2,300 in gold has been taken out in recent years in test runs.

Quartz veins on Blacktail Mountain between these two streams have been mined by open cuts and short adits, none of them longer than 300 feet. The veins on the Tip Top property are numerous, but few of them are as much as 2 feet thick and many are only a fraction of an inch thick. They are commonly parallel with the beds or cut across the beds at low angles. Cross veins perpendicular to the beds are not numerous. The veins pinch and swell and form podlike lenses or become contorted into small folds. In places the quartz ramifies through the wall rock in the form of tiny veins, some of them microscopic in size. The veins are flat-lying or have very gentle eastward dips. They are developed along the outcrops by many openings, none of which are long enough to reveal the extent of the oxidized zone and the changes in mineralogy at depth. The stopes are not large. The quartz vein being mined in any particular adit may be only a foot or two in thickness, but the wall rock above and below the vein may carry gold. One of the richest veins seen in a shallow prospect on the Bondholder claim, on the north side of Blacktail Mountain, is 1 foot thick. This vein contains abundant visible grains of gold.

In addition to gold, the only minerals seen in the veins are quartz, sericite, sphalerite, pyrite, pyrrhotite, and iron oxides. Assays show that small amounts of gold and silver are present in ores where no gold is visible. The sulphides are very scanty and in most places are represented only by their alteration products. Iron oxides form less than 1 percent of most of the ores seen. The grains of gold are irregular in size and shape but are commonly very thin, rough, and spongy or gnarled. Under a hand lens 20 to 30 grains can be seen in some specimens within the space of a square inch. The high-grade quartz is rough and honeycombed where sulphides have weathered out, and in some places it shows ribbon texture, but elsewhere quartz of this same appearance is barren.

The rocks that contain the most productive veins are white sericitic sandstone and light-gray sericitic shale. The sandstone is thin-bedded, and some of the individual beds are very thinly laminated and so weakly cemented that a specimen can be broken in the hand. Where seen near the veins both sandstone and shale contain tiny grains of iron sulphide or iron oxide or holes where these minerals have weathered out. Some of the white sandstones that contain very little metallic mineral but are cut by quartz veinlets a fraction of an inch thick will yield gold upon panning. According to Mr. Miller, these beds contain from $1 to $7 in gold to the ton through a thickness of several feet. That the veins have been faulted along the walls is shown by grooved and slickensided quartz. They have also undergone small vertical displacements along obscure cross fractures that strike northwest and dip southwest. According to Mr. Miller, good ore has been found near these faults, as at the Fisher Creek ("Branagan") mine, across Bramlet Creek from the Tip Top. Microscopic examination of Tip Top ores shows that both gold and sulphide mineralization were related to cracked and brecciated areas in the quartz. Such areas were evidently the more easily permeated by the mineralizing solutions. An assay of a picked specimen, the thin section of which shows a striking relation between sulphide mineralization and shattered quartz, yielded 4.91 ounces of gold and 1.39 ounces of silver to the ton. There was no visible gold in this specimen, and sulphides were rather scanty.

Williams

The Williams group of seven claims is near the head of a valley that lies between Twin Peaks and Great Northern Mountain and is drained by a tributary to Libby Creek. Some of the claims are on top of the ridge that extends westward from Great Northern Mountain to Twin Peaks. The property is reached by trail from the Betty Mae prospect.

Bed veins and crosscutting veins in relatively horizontal or gently dipping gray sandy shales have been opened by several short adits and open cuts between altitudes of 5,900 and 6,400 feet. The longest adit has about 200 feet of workings but no stopes. The open cuts at an altitude of 6,400 feet are on top of the ridge at the head of Standard Creek, and here the beds are horizontal.

The bed veins contain as much as 2 feet of quartz with small amounts of galena, partly altered to lead carbonate and sulphate, and a little iron oxide, some of which is pseudomorphous after pyrite. In some places the veins are not confined to any one bed but grade into lodes of crumpled mineralized shale. Movement has taken place along the walls and within the veins, especially between vein quartz and included fragments and slivers of sericitic shale that have been incompletely replaced. The sericite appears to have acted as a lubricant to aid in faulting.

The crosscutting veins, which strike east or northeast and dip toward the south or southeast at steep angles, are, in some places parallel with the general direction of shearing developed in the shale. They range in thickness from a few inches to 2 feet, and any one vein varies rapidly in dimensions along its strike or dip. These veins are much richer in galena and its oxidation products than the bed veins, and, in addition, a little chalcopyrite partly altered to iron oxide is present. As the exposures of the veins are not numerous, these differences in mineralogy may not be significant.

A selected specimen rich in galena, containing small amounts of chalcopyrite yielded, on assay, 0.04 ounce of gold and 8.71 ounces of silver to the ton.